I SPY
At the BEACH

by Spencer Brinker

Consultant:
Beth Gambro
Reading Specialist
Yorkville, Illinois

Contents

At the Beach................2

Key Words 16

Index......................16

About the Author 16

BEARPORT
PUBLISHING

New York, New York

At the Beach

Where am I?
I am at the beach!

What do I spy?

I spy sand.

It is wet.

I spy water.

It is blue.

I spy a ball.

It is round.
Boing!

I spy a bird.

It is big.

I spy an umbrella.

It is yellow.

I spy children, too!

They are playing.

Key Words

ball

bird

sand

umbrella

water

Index

ball 8–9   sand 4–5   water 6–7
bird 10–11 umbrella 12–13

About the Author

Spencer Brinker lives and works in New York City. In such a big city, you can spy almost anything.

Teaching Tips

Before Reading
- ✔ Guide readers on a "picture walk" through the text by asking them to name the things shown.
- ✔ Discuss book structure by showing children where text will appear consistently on pages.
- ✔ Highlight the supportive pattern of the book. Note the consistent number of sentences found on each page.

During Reading
- ✔ Encourage readers to "read with your finger" and point to each word as it is read. Stop periodically to ask children to point to a specific word in the text.
- ✔ Reading strategies: When encountering unknown words, prompt readers with encouraging cues such as:
  - **Does that word look like a word you already know?**
  - **Check the picture.**

After Reading
- ✔ Write the key words on index cards.
  - **Have readers match them to pictures in the book.**
  - **Have children sort words by category (words that include the letter *b*, for example).**
- ✔ Ask readers to identify their favorite page in the book. Have them read that page aloud.
- ✔ Ask children to write their own sentences. Encourage them to use the same pattern found in the book as a model for their writing.

Credits: Cover, © Anton Gvozdikov/Shutterstock; 2–3, © pkazmierczak/iStock; 4–5, © Ideeone/iStock; 6–7, © ESB Professional/Shutterstock; 8–9, © Fotoschab/Dreamstime; 10–11, © CasarsaGuru/iStock; 12–13, © holgs/iStock; 14–15, © Valua Vitaly/Shutterstock; 16T (L to R), © Fotoschab/Dreamstime and © CasarsaGuru/iStock; 16B (L to R), © Ideeone/iStock, © holgs/iStock, and © ESB Professional/Shutterstock.

Publisher: Kenn Goin **Senior Editor:** Joyce Tavolacci **Creative Director:** Spencer Brinker **Photo Researcher:** Thomas Persano

Library of Congress Cataloging-in-Publication Data in process at time of publication (2019)
Library of Congress Control Number: 2018048656
ISBN-13: 978-1-64280-224-5 (library binding) | ISBN-13: 978-1-64280-397-6 (paperback)

Copyright © 2019 Bearport Publishing Company, Inc. All rights reserved. No part of this publication may be reproduced in whole or in part, stored in any retrieval system, or transmitted in any form or by any means, electronic, mechanical, photocopying, recording, or otherwise, without written permission from the publisher. For more information, write to Bearport Publishing Company, Inc., 45 West 21st Street, Suite 3B, New York, New York 10010. Printed in the United States of America.

10 9 8 7 6 5 4 3 2 1